Introducing the Author

Caroline Bancroft, leading authority on the Tabor story, is just the right person to write about the Matchless mine in Leadville; its history, inextricable from the drama played by Colorado's Silver King and Queen.

Miss Bancroft has an enormous enthusiasm for her native state and has already written two readable publications on this fascinating legend. They are the booklets: *Silver Queen: The Fabulous Story of Baby Doe Tabor* and *Augusta Tabor: Her Side of the Scandal.*

Miss Bancroft's histories are accurate for two reasons. She has spent years in studying and writing about Colorado and she is conscientious in presenting the facts. Her published work about our colorful state has appeared in a variety of different media—in scholarly journals, in slick paper magazines, in newspaper articles, in book reviews, in local magazines, and in book form, notably in the series of Bancroft Booklets designed for tourists (see the back cover) and in two fine full-size books, *Colorful Colorado* and *Gulch of Gold.*

In the accompanying candid photo of 1952, she and a friend are shown picking up ore specimens from one of the dumps on fabulous Fryer Hill. In the background can be seen the ruins of the Matchless mine as it looked before the restoration program was begun a year later.

DAN THORNTON
Governor of Colorado
1951-1955

TABOR $$$$$ IN SHINING SILVER BULLION

* * *

THE MATCHLESS MINE

Although the Matchless in Leadville is probably not Colorado's richest mine, it certainly has been the state's most publicized. Accurate figures for the various mines are not available; but it is known that for years the Matchless poured forth its silver wealth at the rate of $80,000 a month, and that in one or two instances the monthly figure hit as high as $100,000. Its steady shipment of ore to the smelter, where the silver was turned into bullion, was the main sustaining prop of Horace A. W. Tabor's fortune as Silver King of Colorado during the booming 1880's.

The story of how Tabor, a modest storekeeper in the six-months old town of Leadville, grubstaked two prospectors to $17 worth of supplies in May, 1878, for a third interest in their findings, has been told over and over. He later added a hand winch, wheel barrow and other tools to the first grubstake, bringing his total outlay to about $64. What the two prospectors, George Hook and August Rische, found was the Little Pittsburgh. Their claim proved to be a bonanza and established Tabor as a wealthy man. The new millionaire enjoyed the mine's immense returns until September, 1879, when he sold his interest to David Moffat and Jerome Chaffee for a cool million cash.

Tabor looked around for another investment. In all of his other ventures he was involved with partners. For a change he decided he would like to make a purchase for the joy of owning something on his own—to furnish him with a "little pin money," as he jokingly expressed it, "a suit of clothes, a new hat, or a bottle of champagne." He asked for suggestions from the mine speculators, Tim Foley, A. P. Moore and T. B. Wilgus, who had a brokerage company in Leadville.

They offered him the Matchless, an almost unproven claim, but close to the proven Chrysolite and Little Pittsburgh. The Matchless had been found in July, 1878, by Peter Hughes and six other impoverished prospectors, who named it after Lorillard's popular chewing tobacco of that day. This group started sinking a shaft, but they found no ore and sold out. The buyers, Foley, Moore and Wilgus, continued with the shaft until it touched mineral and then waited to make a sale. Horace Tabor proved their buyer. He paid them $117,000 and spent several additional thousands in suits to establish a clear title, making the total price of the Matchless around $150,000.

For years afterward, while other mines played out, the Matchless held up. In 1882, Augusta, Tabor's first wife, valued it in her divorce complaint at $1,000,000. The following year its value was placed at $1,900,000. In Tabor's lifetime the mine was worked profitably from seven different shafts and its total output was said to be close to $7,000,000. Although its production had begun to fall off before silver became worthless in the Panic of '93, it was still a rich mine, listed at close to a half million dollars in value.

What gave the Matchless a special place in history was the role it played in the life of Tabor's second wife, the beauteous "Baby" Doe Tabor. She had been born Lizzie McCourt of Oshkosh and was married first to Harvey Doe, who brought her to Central City, Colorado. This started her on the way to becoming Silver Queen in the '80s, penniless in the '90s, and listener to Tabor's dying words:

"Hang on to the Matchless. It will make millions again."

From 1899 on, Baby Doe did just that. For nearly thirty-six years, she struggled and starved. She sold her jewels, fought legal battles, borrowed money and found leasers, doing everything in her power to hang on to the Matchless. She succeeded. Finally, in March, 1935, when she was eighty years old, her frozen body was found in the Matchless cabin and her long vigil was ended.

After Baby Doe was buried beside Tabor in Mt. Olivet cemetery in Denver, people began to recall anecdotes of the Tabor story and, particularly, of the Matchless. One of the most amusing tales concerned the visit of Oscar Wilde to Leadville in April, 1882, when the Matchless was in its glory.

The celebrated author and wit arrived at the ten-thousand-foot high camp in a spring blizzard after a hard all-day train trip from Denver. He lectured at the Tabor Opera House that evening, despite not feeling well. After the lecture, he was taken to a variety hall and then, on the invitation of Governor Tabor (he was lieutenant-governor but liked the larger title), Wilde was driven to the Matchless between one and two in the morning. A London reporter and Wilde's manager accompanied the lecturer on art and aesthetics.

Wilde was dressed in a slouch hat, corduroy coat, low shoes and tight pants, which the mine superintendent, Charles Pishon, considered inadequate. He gave Governor Tabor's special underground suit to the author to put on. This outfit was made entirely of India rubber and designed for a very tall man which, Eugene Field reported, "having a certain goneness in the length of the pantaloon legs, fitted Wilde quite too."

The lecturer was taken down No. 3 shaft, considered one of the dryest on Fryer Hill but, at best, dirty and cold. Wilde recalled later, in his amusing *Impressions of America*, that he had to descend "in a rickety

THE MATCHLESS NESTLES IN A DRAW ON FRYER HILL

This view was taken in October, 1952, looking northeast in the direction of Mosquito Pass. Where the burros race annually can be seen in the background beyond the long, timbered swell of Prospect Mountain. The gulch, which leads up to the pass, is named Evans and it was down Evans' vaulting length that the stagecoaches careened. There was also a spur of the narrow gauge railroad to serve the mines, running up Evans' lower reaches. A branch came up this draw, its grade still plain below the hill.

bucket in which it was impossible to be graceful. Having got into the heart of the mountain, I had supper, the first course being whisky, the second whisky and the third whisky."

The *Denver Tribune* said he drank twelve snorters, and that the miners voted him a perfect gentleman. Pishon showed him about, pointing out rich chlorides and carbonates. Wilde was disappointed that they were so ugly and sombre-looking. In the gloom, he gathered the folds of the rubber suit more carefully about him and remarked:

"This cloak reminds me of the togas worn by the Roman senators. The lining, however, should be of purple satin with storks and fern embroidery."

The local press took umbrage and remonstrated pompously:

"The aesthete seemed to forget that the toga was one of an American Senator and not of an effete Latin statesman."

But Wilde's loquacity and humor were not lost entirely in the Cloud City. At the time, Leadville had the reputation for being one of the richest cities in the world, and also the roughest.

"Nearly every man carried a revolver," Wilde recollected. "I was told they would be sure to shoot me or my traveling manager. I wrote and told them that nothing they could do to my traveling manager would intimidate me."

Wilde also said that in the variety hall he had seen a sign. True or not, Leadville has been quoting it ever since. The sign read:

> *Please do not shoot the pianist.*
> *He is doing his best.*

While he was down in the Matchless and stimulated by the repeated libations, the author practiced several of these witticisms on the miners. But the men apparently took a dim view of his sallies, for the *Rocky Mountain News* said that "he acted like a lunatic while down below."

Still, it is not every mine that has harbored in its dark intestines such a rakish scene played by an international celebrity. The Matchless is unique in this regard as well as in the long-continuing drama that it played in the lives of the Tabors.

It is also unique in the fact that in its later years, not one, but two famous Colorado millionaires redeemed it for Baby Doe out of sympathy for her cause. One was W. S. Stratton who, in 1901, gave her a check for $14,858 to pay off a judgment against the mine. But when a man, who had a judgment against the Tabor Grand Theatre in Denver, established a prior lien against the Matchless, Baby Doe returned the check to the Gold King of Cripple Creek.

In July of that same year, the mine was sold at a sheriff's sale for $3,866. Claudia McCourt of Chicago, Baby Doe's favorite sister, got it

FABULOUS FRYER HILL FROM FIRST TO LAST

The only similarity in the seventy years' difference of these two photos is the long slope of Prospect Mountain and the mine dump high on its brow. W. H. Jackson took the earlier view in the '80s when Fryer Hill was populated with mines, stores, residences and saloons. The big mine with the long trestle for dumping ore is Tabor's Little Pittsburgh, his first lucky strike. The Matchless shows only with its roofs, hidden behind the top of the tall, dark pine tree. The near road is an extension of Leadville's Seventh Street; the far road, of Eighth Street. Below is the Matchless in ruins, prior to its 1953 partial restoration and present use as a museum.

back for the distressed widow. Baby Doe moved from Denver to Leadville, fixed up a tool cabin close to the hoisthouse as a dwelling place, and made the Matchless her permanent residence.

In the succeeding years the mine's history was one of litigation, difficulty with leases, and indifferent returns. Its only continuous operation was from 1908 to 1911 under H. J. Stephens who was mining ore worth about $10 a ton.

By 1928, Baby Doe was in serious difficulty again. It was reported that "The Unsinkable" Mrs. Brown was about to produce the necessary $14,000 and redeem the mine for Mrs. Tabor. But it was J. K. Mullen, millionaire miller, who paid off the mortgage. Simultaneously he created the Shorego Investment Co. which protected Mrs. Tabor in her ownership until her death. This company is still the owner of the mine although management has devolved on the Mullen heirs.

In the years immediately after Baby Doe's death, leasers again worked the mine. But the profits were too lean, and operation was shut down. Gradually the mine fell in ruins, the machinery was disposed of, and Baby Doe's personal cabin was vandalized by souvenir hunters. By 1952 only a shell of the mine's former splendor remained. It stood stark and alone on Fryer Hill, almost like a ghost, a tragic reminder of Colorado's spectacular past. Once again its plight demanded help.

The Leadville Assembly, Inc., a non-profit organization, was formed to attempt a restoration of the small dwelling cabin at the mine and perhaps in succeeding years, if public support proved adequate, to make more extensive repairs. The organization began its work in 1953 and dedicated the Baby Doe Tabor cabin as a museum on July 16 under the leadership of Caroline Bancroft, Colorado historian and Tabor authority.

The Leadville Assembly is still in need of funds to carry out future restoration of the shaft and hoisthouse. But in the intervening years, the organization has made the Matchless Mine Museum a historic monument where visitors can stand on breath-taking Fryer Hill and savor the atmosphere of the past—a boisterous, bonanza past when the Matchless was just one of many noisy mines pouring out riches for Colorado's lustiest camp—Leadville.

For full and accurate details of the Tabor story, the reader is referred to "Silver Queen: The Fabulous Story of Baby Doe Tabor," whose charm and liveliness have gained the booklet a wide audience.

LUSTY LEADVILLE

Leadville is renowned in Colorado as the state's richest silver camp. (It was also probably the rowdiest but of that, more later.)

Yet oddly enough the town began and ended its nineteenth-century career as a gold camp—not a silver camp.

It happened this way:

In 1859 the Colorado Rockies became the target for the nation's second big gold rush (the first having been to California ten years earlier). Spectacular finds were made along the creek beds close to the present sites of Boulder, Central City and Idaho Springs. Ramshackle towns immediately sprang up close to the gold placers. As thousands upon thousands poured across the plains and swarmed into the mountains, the later arrivals found that the first-comers had staked all likely claims in these localities.

So they swarmed on—up the Platte, over Kenosha Pass and into South Park; up the continental divide, over Hoosier Pass and down the valley of the Blue; or over Trout Creek Pass and up the Arkansas river bed. Then howling winter set in, and the frightened gold-seekers fled the high snow-blanketed mountains and ceased their avid searching.

With the first thaws of spring they were back. By April, 1860, a number of parties were camped along the Arkansas. One of these groups included Horace and Augusta Tabor who were trying their luck close to the present town of Granite. The men in the Tabor Party were becoming discouraged with their placer because wherever they panned, they encountered heavy particles of black sand mixed in the gold. This sand was extremely difficult to separate from the gold and radically cut down on profits.

Higher up on the Arkansas, other parties of men were panning the gulches that ran into the valley from Mount Massive on the west, while still others worked the gulches cutting down from the Mosquito Range on the east. It was in one of these that Abe Lee straightened up on April 27 and hallooed to the other members of a party led by George Stevens.

"O my God!" Lee yelled, "I've got all California in this here pan."

And so the gulch was called California. It runs down off the southwest side of Ball Mountain (one of the lower elevations of the Mosquito Range) and enters the Arkansas in the region of the present town of Malta.

As soon as the news was out, the Tabor party struck camp and headed toward California Gulch where they arrived May 8, 1860. Augusta Tabor was the first woman there, and the men promptly built her a primitive log cabin with a sod roof. The exact location is unknown; but she spoke of it as being in the "upper end of the gulch." She did laundry, took boarders, weighed gold dust, and in the intervening weeks when ten thousand men crowded into the gulch's confines, Mrs. Tabor was appointed postmistress.

Various huddles of wagons, tents, brush huts and log cabins quickly grew up along the seven-mile-length of the gulch (which runs parallel to the southern boundary of present day Leadville) and this long string of habitations became known as Oro City.

The settlement grew rapidly since most of the claims proved extremely rich.

No accurate figures exist for those days, but it has been assumed that some $2,000,000 worth of gold was taken out that summer. Claims took up every foot of the gulch from its source down to the spot where it joined the Arkansas. Each day men were working hard with gold pans, rockers, Long Toms and sluice boxes from early dawn until the light failed. Oro City's life was one of frenzied activity.

Oddly enough, the claims varied drastically in riches; and this was not entirely due to the skill of the worker in separating the gold from the black sand. Tabor's claim netted him $5,000 that summer while that of the man just below him on the creek bed brought in $80,000. This variance was probably due to the erratic strength of the trickle of water in the stream bed which had been depositing gold in unpredictable pockets during its centruies-long flow.

Oro City was soon the largest settlement in the land that now constitutes our state. (Colorado Territory was not created until February 28, 1861.) The town's character was typical of any early-day mining camp—ramshackle, rough, dirty, boisterous and devil-may-care, but remarkably honest. There was almost no law or order except that of mutual consent arrived at by the founders of the mining district.

Abe Lee was appointed recorder, and it was agreed that jumping a claim, stealing and murder were not to be tolerated. Punishment was summary and drastic. Despite the fact that tents and cabins had no doors, there was almost no thievery.

Other vices abounded. Only a very few wives accompanied their gold-seeking husbands; so Oro City men took their pleasure and ribald companionship where they found it—in the gambling dens, dance halls and sporting houses. These sprang up with the same rapidity as habitations and ordinary places of business. One sporting house was an enormous affair, built solely of pine boughs, while another did business in a stalwart log cabin with a dirt floor. The story goes that the night before four miners were to leave California Gulch, they scraped the surface dirt from this floor, mindful that many a man had paid his light-of-love in gold dust. The miners ran the dirt through their sluice box and netted more than two thousand dollars.

The most colorful character of California Gulch's underworld was a mysterious twenty-year-old girl. Her charm and cultivated Boston accent captivated all the better class of men in the settlement; while her prettiness and habit of wearing red ribbons in her dark hair complimented by flashing red stockings over her trim ankles attracted the others. She called herself Nellie but never told her last name. The girl arrived in camp in June with an eye-catching wardrobe, obviously made by the most fashionable Eastern modistes. Every day she rode about camp on a spirited horse, displaying her figure to advantage. At night she received callers in a log cabin that she had bought from a miner. It was not long before she was known the length and tiny breadth of the gulch as Red Stockings.

Red Stockings told her story to a confidant or two. Born to wealth and position, three years previously she had been given a trip to Europe in a party of four. There she met a young French officer with whom she entered into a clandestine affair. Despite her repentance on returning home, she was never received again by the society she had known. So she ran away with a gambler with whom she stayed until he started to drink too heavily. Then she elected to seek her fortune alone in the gold camps and chose California Gulch.

The gulch recorded only one murder that summer. With the aid of a shotgun a man by the name of Kennedy tried to maintain an illegal hold on a mining claim. When he attempted violence toward the rightful claimants, one of them with a rifle was faster on the draw. Kennedy was buried, and a miners' meeting was held to decide what to do with the young owner of the rifle. He was acquitted.

The rich exciting life of Oro City lasted but two summers. By the autumn of 1861 most of the claims were played out. Placer gold was recoverable in only infinitesimal amounts, and the dwellers of Oro City faced this fact unsentimentally. They deserted their cabins and sought an easier life elsewhere. The Tabors moved across Mosquito Range to the mining camp of Buckskin Joe. Red Stockings disappeared. But it was rumored that she left with a hundred thousand dollars, immediately reformed, and became a fine wife and mother in Nevada.

Only the hardiest souls remained through the next years and by 1865 were estimated to be less than four hundred. The number of inhabitants continued to diminish in the face of fabulous discoveries in Montana and Idaho, and by 1868 Oro City was hardly more than a ghost town.

In that year the Printer Boy lode, a gold mine on the south side of upper California gulch, hit a profitable streak. It had been discovered in 1861 but never adequately worked. Now the mine's solid production gave new hope to the remaining inhabitants and encouraged a few old-timers to return. Among the latter were the Tabors.

The Oro City activities now clustered close to the Printer Boy. The Tabors opened a store, nearly two miles up the gulch from the southern end of today's Harrison Avenue. A group of buildings (now in 1960 mostly gone) about two and a half miles above the same spot was officially desig-

HOME LIFE IN 1879

The "boomers" who rushed to find fabulous silver in the lead carbonates around Leadville had many hardships to endure. They lived in tents and primitive log cabins and spent most of their money for mining tools. Tots (like the one shown) often did not survive. Colorado mountain cemeteries are filled with graves of babies and children.

nated as Oro. The gulch's inhabitants looked forward to a re-birth of their town.

But this hope was short-lived. The Printer Boy's production fell off, and by 1870 Oro was once again a candidate for ghost-town status. The Tabors still clung to the operation of their store, but they were alone in commercial activity. For several years there was not even a saloon to keep them company.

California Gulch slept on, visited occasionally by hunting and exploration parties and harboring a few die-hard placer miners who continued to curse the heavy black sand and rocks that clogged their sluice boxes and hindered their digging. Only two men were curious about this black rock. They were William H. Stevens and Alvinus B. Wood. Like all the others, their original aim was to mine gold. They thought if they could obtain adequate water by bringing a flow from the Arkansas via a twelve-mile ditch, the problem of the black rock would be solved by hydraulic pressure.

They began operation in 1874, and their company earned some twenty to thirty per cent on their original investment of $50,000. Commercially their venture was a success; but the hydraulic pressure did not overcome the handicap of the black rock. In June they took samples of this rock from the south side of the gulch about a mile above present Leadville and sent them to be assayed. The reports proved the black rock to be silver-lead carbonates running from twenty to forty dollars a ton in silver alone.

Stevens and Wood kept the knowledge of silver-bearing ore a secret for two years. They freely admitted the black rock contained lead but added nothing more. After studying the geology of the region, they staked claims that almost covered both Rock and Iron Hills. The partners started mining seriously in the summer of 1876. But after they had paid the cost of ore haulage via ox-team to Colorado Springs and via rail to the St. Louis smelter, their accounts were in the red. Wood decided to sell out.

LUSTY LEADVILLE OFFERS THE TOURIST MANY INTERESTING AND

THE TABORS MADE BOTH OF THESE FAMOUS

Above is the Matchless Mine as it looked after Baby Doe's cabin was turned into a summer museum. Each year the site draws thousands of visitors. Below is the Vendome Hotel (taken by W. H. Jackson in 1884) shortly after being completed. At that time the hotel was named the Tabor Grand after Horace Tabor, who financed its completion.

RIC BUILDINGS FOR FASCINATING AND NOSTALGIC SIGHT-SEEING

MANY ODD RESIDENCES WERE IN VOGUE

Above are two homes which are now owned by the Colorado Historical Society and are open to the public. The Healy House (left) was characteristic of 1879 mansions, and the Dexter Cabin was the unique cottage of a wealthy bachelor. Below is the private House-with-the Eye, built in the '80s by Eugene Robitaille after the State Seal.

St. Louis capitalists were interested. The shipments of ore were impressive enough to warrant further exploitation even if the first shipments had proved unprofitable. In 1877 Levi Z. Leiter, business associate of Marshall Field of Chicago, became Stevens' new partner, and Augustus Meyer and Edwin Harrison established a smelting and reduction works close by the carbonate deposits. Development of lead-silver mines began in earnest with big profits promised shortly.

Violence and litigation sprang up in equal earnest. Stevens and Leiter had to fight opposing miners with sulphur smoke below ground and in the courts with legal brains for two years afterward. Eventually they were victorious on both battle fields and established their right to continue mining along the vein.

The summer of 1877 brought prospectors back to the region, this time in search of silver. Stores opened along a street that ran parallel to California Gulch on its north side. This was called Chestnut Street. Tabor moved his store down from Oro City to a location on Chestnut just a lot west from the corner of Harrison Avenue. Here Tabor also housed the post office and an informal bank in his iron safe.

By the end of the year there was evidence of a small boom. Lake County's output for that year, 1877, was the fourth largest in the state with a $670,600 yield in gold, silver and lead. The population had risen to around four hundred, perhaps higher.

The inhabitants, however many, had centered their camp lower than the second Oro City and higher than California Gulch. They felt it had an entity of its own and deserved a new name. On the evening of January 14, 1878, a group of eighteen citizens met and selected the name of Leadville, honoring the lead-silver carbonates that had brought the new camp into being. Soon a town government was set up, and Tabor was chosen mayor. The town's nickname was "Cloud City."

By spring the settlement was in a ferment. Newcomers were pouring in from all directions. Freighting and stagecoach travel were overloaded since the nearest railroads were too distant to be of any help. The South Park railroad had penetrated the Platte Canyon but a short distance and by the latter part of the year was only as far as Bailey, fifty-six miles from Denver. The Denver and Rio Grande was far down the Arkansas at Florence.

The boomers poured into Denver from the East, determined to reach Leadville by any means, which frequently meant walking and scrounging for food as best they could. Sleeping accommodations in the new carbonate camp were at a premium. R. G. Dill, who wrote a history of Lake County in 1881, said:

"For the privilege of lying on a dirty mattress, laid upon the floor of a boarding tent, with a suspicious looking blanket for a cover, and its chances of proximity to a thief or desperado, those who could afford it paid a dollar. Those of lower financial grade were glad to get accommoda-

SUMMER OF 1884

This photograph was taken on August 31, 1884, from Capitol Hill and shows the rear of the Court House on Harrison Avenue, topped by the tower with a snowy figure of Justice. The rigors of the climate were only a small part of the hard battles fought by our intrepid Colorado pioneers in their craze for speedy gold or silver riches.

tions in the dirty sawdust on the floor of a saloon or gambling-hall. In every direction the sound of the saw and the hammer were incessant. Night and day men were employed, at enormous wage, to erect shelters for those who daily thronged into camp. One street—Chestnut—comprised the town, and along this street were packed before the end of summer not fewer than 6000 men. From daylight to the return of daylight again the street was thronged with pedestrians and freighting teams. The latter sometimes blocked Chestnut for its entire length, which occasions were notable for the ingenious oaths of the teamsters and the pistol-like cracks of their bull whips . . ."

By the end of summer a parallel street, State (now Second), had been built on the north. It soon attracted a majority of the lower element devoted to drinking, gambling, variety halls, and sporting houses although such activities were by no means confined to this one area. Leadville drew an unusually large percentage of the vicious. Footpads, thieves, murderers, confidence men, harlots and even members of the infamous Jesse James gang thronged the streets or hid in outlying gulches such as Half Moon Gulch. During this year and the next not all the quick and easy riches were obtained from ore. The gullible and defenseless also yielded up their quota.

Respectable businesses, religion and education also made appearances in 1878 and obtained firm footholds. David May (who arrived in Leadville the autumn before and opened a small store housed in a tent) built a frame building with a false front at the beginning of the year and was soon prospering. After a ten year residence, twice serving as Lake County treasurer, he moved to Denver and started the foundation of the May Company chain of stores. Thomas Uzzell, a young Methodist minister, started holding services in a log cabin February 1. His sermons attracted an enthusiastic following who built a church soon after. Other denomina-

CHESTNUT STREET

A traffic jam was forcing an ore wagon onto the boardwalk when this photo was taken in 1880. Note the large silver dollars exhibited atop the two buildings to the left —Tabor's new bank and his old store. He loved the idea of a silver dollar so much he named his second daughter thus. One of these silver dollars may be seen in the Silver Dollar Grill.

tions formed congregations, and religion was established. School began that spring with about thirty children meeting in a log cabin. But with the constant influx of families, the school was enlarged by fall to a small frame building accommodating sixty pupils and two teachers.

The excitement and growing population kept up for the next two years. Every sort of melodramatic event occurred in such quick succession that the usual TV serial would be shamed. Murders, duels, mine-saltings, mine-feuds, stagecoach accidents, horse runaways (sometimes hilarious), foot-pad violences, fist fights, gang brawls, suicides of "soiled doves," lot-jumping, mysterious disappearances, titled tourists, and new millionaires were the order of the day in the columns of Leadville newspapers.

New hotels sprang up. The first notable one was the Grand Hotel on Chestnut Street run by Thomas F. Walsh (later to make millions in Ouray and become an intimate of kings and presidents). The second was the Clarendon on Harrison Avenue, erected by William H. Bush who had been proprietor of the Teller House in Central City until January, 1879. He and his brother, James, a livery-stable owner, moved to Leadville that month. They bought some lots and staked claims to others from which they were able to make fantastic profits. One known lot worth $50 in 1877 sold for $500 the next year and by 1880 changed hands at the price of $7,500. This phenomenal rise was true for all good business locations.

In March the Bush brothers sold one of their lots to the partner of a lumber company. Before the new owners could take possession, the lot had been jumped by Mortimer Arbuckle who erected a slab shanty and a fence on it one morning before breakfast. An altercation ensued between Arbuckle and the Bush brothers which led to a scuffle between Arbuckle and Bill Bush. Suddenly a shot ended the scuffle, and Arbuckle lay dead from a bullet fired by Jim Bush's pistol.

The hot-headed brother claimed he had fired to protect Bill. Jim was taken into custody, and a hearing was held that afternoon. The temper

HARRISON AVENUE

After Bush built the Clarendon and Tabor erected the Opera House, both in 1879, the trend of the city started northward. Soon Harrison Avenue had completely usurped the old place of Chestnut. This 1884 photo shows the new Court House with Justice aloft and barely-completed Tabor Grand Hotel (now the Vendome) with its fashionable mansard roof.

of the camp was very ugly, partly because lot-jumping violences had reached a scandalous state and partly because Arbuckle was unarmed. There was wilder and wilder talk of lynching Jim or burning Bill's nearly-completed hotel. A special guard of one hundred men was organized, and just before dawn Jim was surreptitiously taken from the jail to Denver for safe-keeping. Later the case dragged on for many years until acquittal was finally obtained, probably through judicious bribery.

The Clarendon, Bill Bush's hotel, opened on April 10 with a deputation of prominent men from across the nation—U. S. senators, mining men and capitalists. Despite the fact that the frame hotel's location, on the northeast corner of Third and Harrison, was considered disastrously distant from the town's main thoroughfare, Chestnut, the Clarendon immediately prospered. Later in August, 1879, Bush and Horace Tabor erected the Tabor Opera House just north of the hotel, using brick construction and connecting the two buildings with a covered passageway over an intervening lane.

The opera house was opened in November. Harrison Avenue was now established as the principal site for substantial building, and there was constant noise of sawing, pounding and hammering along its length. The population increased so materially that Leadville became second only to Denver in size, and Harrison Avenue soon took on the aspects of a boulevard. An 1879 newspaper article listed Leadville as having "19 hotels, 41 lodging homes, 82 drinking saloons, 38 restaurants, 13 wholesale liquor houses . . . 10 lumber yards, 7 smelting and reduction works, 2 sampling works for testing ores, 12 blacksmith shops, 6 livery stables, 3 undertakers, 21 gambling houses ('where all sorts of games are played as openly as the Sunday School sermon is conducted'), 4 theatres, 4 dance halls and 35 houses of prostitution."

Both the commercial activity and the mines attracted many names famous then or later in state and national history. Marshall Field of

Chicago invested in the Chrysolite mine on Fryer Hill. (This was a supposedly spurious mine that Tabor bought of "Chicken Bill" Lovell who had salted it with ore from Tabor's own bonanza, the Little Pittsburgh. But its joke turned out to be on the crook—the Chrysolite developed into an even richer bonanza than the first mine!) Meyer Guggenheim arrived to invest in the A.Y. and Minnie mines, but soon switched to smelting. From this modest Leadville beginning he and his sons developed the now international firm of the American Smelting and Refining Company.

Charles Boettcher was another 1879 arrival. He opened a hardware store on Harrison Avenue across from the Clarendon and prospered so mightily that, after he moved to Denver many years later, he was able to be the main force behind two of Colorado's greatest industries, the Ideal Cement Company and the Great Western Sugar Company. He lived to be ninety-six, and at his death in 1948 the Boettcher fortune, including the additions made by his son, Claude, was likely the state's largest.

Many other names prominent in Colorado history, then or later, were associated with early Leadville. Some were residents and some were investors. All were attracted by the fact that in 1879 Lake County produced over eleven million dollars in ore, some four million odd dollars in excess of all other mineral counties of Colorado taken together. A partial list of these important men would include Governor John L. Routt, Senator Jerome B. Chaffee, David H. Moffat, John F. Campion, Samuel Nicholson (later a United States senator), William H. James, Eben Smith, A. V. Hunter, Thomas F. Daly (founder of the Daly Insurance Company of Denver), John A. Ewing, Max Boehmer, James B. Grant (later governor), James J. Brown (whose wife, Maggie, achieved fame and notoriety as "The Unsinkable Mrs. Brown," survivor of the *Titanic* disaster). Jim Brown developed a gold belt in the Little Jonny mine which had previously been shipping silver ore. By this move he made a million dollars for himself and added to the renown of the Little Jonny after the price of silver slid in the Panic of 1893.

But in 1880 those dark days were thirteen years away. Leadville at the time was a mountain metropolis with everything booming. The official U. S. census listed the population as 14,820; but the Leadville directory for 1880 gave it as over 31,000. Undoubtedly the first tally confined its count to the city limits while the second included all the nearby settlements such as Oro City, Adelaide, Finntown, Evansville, Stumptown and the like. These were suburban mining settlements of one hundred to a thousand people although today they are ghost-towns. The Leadville directory probably also included the many restless boomers and prospectors who moved in and out of camp and were not resident long enough to be found home by the census taker. It seems safe to say that in 1880 Leadville numbered over 20,000 more or less stable population.

A RED-LIGHT DISTRICT FOLLOWED COMMERCE

The first street was Chestnut (below) which ran east and west parallel to California Gulch. This view looks east toward Carbonate Hill and shows Thomas F. Walsh's Grand Hotel on the south side with Tabor's store and bank farther on at the corner. State Street (above) was next on the north, running parallel, and was crowded with variety halls (Comique and Odeon), cribs, saloons and clip joints. This photo looks west toward Mt. Massive and gives a hint of Leadville's spectacular setting, facing the high snow-topped Continental Divide.

ORNATE FUNERAL

The Tabor Light Cavalry lent style and dash to events of the early '80s. Their unusual uniforms of red trousers, blue coats and brass helmets and their flashing sabers were imported from New York as the gift of Tabor. They were organized during the strike of 1880 but did not disband after the crisis. A member's death called forth real pomp.

This same year also brought increasing absentee-ownership of the mines and over-capitalization. In May a serious strike of the miners resulted in Leadville's being put under martial law during the middle of June by Governor Frederick W. Pitkin. The miners struck for an eight-hour day and a minimum wage of $4 a shift. Angry, the managers for Eastern companies stood firmly against these demands, and local public sentiment backed up the managers. After the creation of armed units such as the Tabor Light Cavalry, interspersed with some violence, a number of parades, and much heated talk, the strike was lost.

The history of Leadville during the decade of the 1880's was one of continued heavy production. The high figures began in January when one of the richest mines on Fryer Hill, the Robert E. Lee, set out to make a record. During a seventeen-hour stretch of January 4, some $118,500 was extracted. This record was never equalled by any other mine in Colorado, although its neighbor to the north, Tabor's Matchless, proved to have enormous staying powers of wealth. The Matchless began producing in September of that year and never played out while there was a demand for silver. These mines were only two of many that made the figures for Lake County production jump from $11,285,278 in 1879 to nearly fifteen million dollars in 1880.

From 1879 until 1892 Lake was the richest mining county in Colorado, despite the fact that its figures fell gradually from the 1880 high peak to $7,856,000 in the year of 1892 when it was topped by Aspen's production for Pitkin County. During this period its business remained solid. In 1893, the year of the Silver Panic, Leadville regained first place with a production of over $8,000,000. After the government stopped buying silver, gold was king again, and the phenomenal rise of Cripple Creek topped all. Yet Leadville held fast to second place for many years.

The Cloud City's course after the Silver Panic was winding. A successful effort was made to develop the small gold belt on Breece Hill, and

BIZARRE BANQUET

Seldom is a party held in the bowels of the earth and reached through the drifts of a damp dark mine. But the discovery of zinc's worth in January, 1911, produced such an occasion. Here, some of the guests of Samuel Nicholson, who were waiting to descend the shaft of the Wolftone Mine, pose for their photos. Below a hot Scotch punch awaited them.

some of the silver mines continued producing for slim returns. In the autumn of 1895 an effort was made to attract tourists who might bolster up the sagging economy. An enormous palace was constructed entirely of ice blocks. The whole venture was typical of the exuberance and expansiveness of this lusty mining camp.

A grand opening was held on New Year's Day, 1896, and the building became the wonder of the state. Unfortunately a March thaw, very early for Leadville's usual climate, meant that the Ice Palace had to be closed after only ten weeks' use. Not long afterward it evaporated completely. Still, everyone agreed the project had been a lot of fun even if it was a financial failure.

During the remaining 1890's and early 1900's Leadville's decline was steady. More and more people moved away. The most popular choices for new residence were Denver or Cripple Creek, and Leadville slid into the realm of the forgotten.

Then in 1910 the value of zinc was discovered. Previously many of Leadville's mines had been throwing out their zinc carbonates as waste. But increasing industrialization and World War I created a heavy demand for zinc, and also lead. Leadville experienced a new roaring time.

The Wolftone, the Mikado and the Penrose were a few of the mines that achieved a lively renascence. To celebrate this happy discovery, a banquet was held in 1911 on January 25, birthday of Robert Burns. Two hundred and fifty people gathered in a glittering hundred-foot stope of the Wolftone, over a thousand feet below ground. The mayor had proclaimed the day a legal holiday, and every sleigh and transfer wagon in town was pressed into service to transport guests to the mid-day festivity. People arrived from all over the state and from as far away as Chicago and Milwaukee. Scottish bagpipe players, an orchestra and a hot Scotch punch enlivened the occasion as well as the many Scotch and Irish jokes interlarded in the speeches. Happy days were here again.

The demand for zinc and lead lasted until after World War I, then Leadville again experienced hard times. It was during the 1920's and 1930's that the populace reverted to its former lawlessness. Open gambling, in defiance of a state ban, and the operation of many stills for producing bootleg "mountain dew," a sugar whisky, were widely accepted for aiding the exchequer. State Street had deteriorated from its former infamous glory to a few saloons and a row of cribs. But it was prepared to entertain visiting hunters and fishermen in the manner of the old West. With these meagre pickings the town limped along.

Simultaneously operations were beginning on Bartlett mountain, some twelve miles to the north, that were destined to change the whole story of Leadville. Much earlier, close to Fremont Pass an enormous deposit of molybdenum, a rare mineral used in the hardening of steel, had been discovered. Yet it was 1924 before the commercial possibilities of molybdenum at Climax seemed feasible. Leadville was not much interested at the time. Most of the residents could not pronounce the name of the mineral, and the Climax Mining Company's production figures and profit for 1925 were too modest for comment.

The passage of years was to alter all that in the view of Leadvillites. Their own lead and zinc deposits, augmented by gold and silver, staged a comeback in the 1930's which proved profitable through World War II and in the post-war years. But by the half-century mark comparative figures were revealing a new balance of power in Lake County. Lead and zinc were falling, and in 1953 the big operators, Resurrection Mining Company (a subsidiary of Numont) and the American Smelting and Refining Company began to pull out. In the succeeding years the small mines shut down, one by one, and the Arkansas Valley Smelter, lacking ore to treat, could be open only on a part-time basis.

In 1954 Leadville's production in round figures looked like this: gold, $3,325,000; silver, $3,077,172; lead, $4,814,400; zinc, $7,714,500, while Climax's figures looked like this: molybdenum, $45,192,856, plus the three minerals found in association with molybdenum: tungsten, $5,577,495; copper, $2,613,600; pyrites, $132,850. Leadville was responsible for less than twenty million dollars and Climax was responsible for more than fifty-three million dollars.

In 1958 Leadville produced $100 worth of gold; $12 of silver; $78 of lead, and $4 of zinc. With less than $200 production of ore, no one could call Leadville a mining town any longer. At the same time Climax's figures ran close to thirty-five million dollars. Seeking jobs, Leadville miners turned increasingly to Climax, and commuting between the two towns became an accepted rule.

Leadville's character inevitably began to change. The beginning of this transition period dawned on November 30, 1951, when the row of frame cribs on State Street burned to the ground. The Pioneer Saloon,

THE CLOUD CITY BRED AN ICY CHIMERA

Webster defines a chimera as a "foolish fancy." Certainly the Ice Palace of 1895-96 (which lasted only ten weeks) was just that. But it was all part of the headiness that attacks people at 10,000 feet altitude. Stock was sold for the venture at a dollar a share and more than $20,000 was raised before construction was begun, only to have it all evaporate! The Cloud City's lofty location is below with Mt. Massive in the background; the Annunciation spire rises in the left foreground and farther back is the Arkansas Valley smelter stack with smoke drifting down river. Smelter is now dismantled and closed.

although damaged, was saved and for a time continued as a stronghold for Leadville's lawless element. But the old lustiness could not survive without mining profits, and the residents had to seek other means of livelihood.

The Cloud City's spectacular setting and many historic buildings had always offered a great tourist potential. Yet for years the residents had snubbed any such idea while the mines produced. Now gradually the town's leaders were won over to a more lenient view of visitors.

Today the Healy House and Dexter Cabin (operated by the Colorado Historical Society), the Matchless Mine Museum, the Augusta Tabor House, St. George's Episcopal Church, the Carriage House, and the Tabor Opera House are open in the summer as tourist attractions (with occasional exceptions dependent on the guide situation). The Vendome Hotel (formerly the Tabor Grand) has been renovated and caters to both summer and winter visitors, the latter drawn to the ski area developed on Cooper Hill, ten miles northwest of town. The Leadville-Fairplay burro race, held annually in mid-summer across Mosquito Pass, has become a nationally known event, and breath-taking jeep tours are conducted to Leadville's formerly big-producing mines, now scattered like ghosts on the Cloud City's outlying hills.

The mines' history was unique. Although their final one-hundred-year production figures, from 1860 to 1960, include only probable amounts for the early California Gulch placers, it is estimated that their total production was around five hundred and fifteen million dollars. This amount was formerly considered fantastic. On the other hand Climax's final figures for a thirty-five-year operating period, from 1925 to 1960, were six hundred and forty-five million dollars—one hundred and thirty million more for around a third as long an operating period.

So Lusty Leadville, which began its formal existence as a suburb of Oro City in 1877, was not much more by 1957, eighty years later, than a suburb once again. To offset this trend, warmth and cordiality toward strangers developed a greater impetus as Leadville sought an economic identity of its own. Even if it meant succumbing to the tourist-town idea, Leadville wanted to survive on its own.

In 1960 the "mining-camp die-hard" adherents achieved a triumph. The Climax Mining Company moved nearly all their housing units down to Leadville, and Climax virtually ceased as a town, becoming merely a mine and a mill. The mining faction of Leadvillites then proclaimed victory.

But the trend had gone too far. Restaurateurs and motel owners wanted visitors and pleaded for a receptive attitude toward tourists. So Leadville continues with a split personality. In this way the town is similar to its setting amongst magnificent mountains which have always communicated an odd combination of idealistic inspiration and physical lustiness. Cloud City still vies with Lusty Leadville.